Do you ever look up at the night sky and see all the bright stars and wonder what's up there? Well, space is what's up there! It's a giant place that's outside of our planet Earth. Imagine a really, really big room that you can't see the walls or ceiling of - that's what space is like.

Space exploration has been a subject of fascination for humanity for centuries. From the first telescopes used to observe the stars to the modern spacecraft that roam the solar system and beyond, we have always been curious about what lies beyond our planet. Space is a vast and mysterious place that has captured the imaginations of people of all ages, especially children. There's something truly captivating about the idea of exploring the unknown and discovering new worlds, and this sense of wonder and curiosity is something that can inspire children to learn and explore the universe. In this book, we will explore some fascinating facts about space that are perfect for children who are curious about the cosmos. From planets and stars to black holes and galaxies, we'll delve into some of the most exciting and interesting aspects of space that are sure to spark a child's imagination.

In space, there are lots of things to see like planets, stars, moons, and even comets! Planets are big balls of rock or gas that orbit, or go around, a star. Earth is a planet and so are Mars and Jupiter. Stars are like big balls of fire that give off heat and light. They come in different colours and sizes. Some are really big and some are small. Our Sun is a star and it's the closest one to us.

Space also has moons, which are like small planets that orbit around a larger planet. Earth has one moon, but some planets like Jupiter have many! Comets are like dirty snowballs that travel through space. They have a tail of gas and dust that makes them look really pretty when we see them from Earth.

One thing that's really different about space compared to Earth is that there's no air or gravity up there. That means things float instead of falling down, which is really fun to imagine! But it also means that people have to wear special suits and helmets to breathe and stay safe when they travel to space.

So, space is a really big and interesting place with lots of different things to explore and learn about. By studying space, we can learn more about the universe and the things that are outside of our planet. It's really exciting and we never know what we might discover next! Space is an incredibly fascinating and mysterious place that has captivated human beings for centuries. As a child, you might have wondered about what space is and what lies beyond our planet Earth. In this book, we'll take a journey through the universe to explore some of the most exciting and interesting facts about space that will surely ignite your curiosity.

Firstly, let's talk about what space is. Space is the vast and empty expanse that surrounds our planet, stars, galaxies, and all celestial bodies. It's also known as the universe and is composed of countless planets, stars, asteroids, and other cosmic entities. Space is a vast, dark, and cold place that can be incredibly beautiful and awe-inspiring.

By learning about space, we can understand more about the universe and the things that are outside of our planet. It's an exciting place full of mysteries waiting to be explored!

One of the most exciting things about space is the variety of celestial bodies that exist. There are eight planets in our solar system, including Earth, which is the only planet we know of that has life. Each planet has its unique features, such as the thick clouds of Venus or the red surface of Mars. There are also dwarf planets like Pluto and Ceres, and comets and asteroids that orbit the sun.

Another exciting aspect of space is the stars. There are billions of stars in the universe, and each star is unique. Some stars are small and faint, while others are massive and bright. The sun is a star, and it's the closest star to Earth. Stars are born from clouds of gas and dust, and they produce energy by fusing hydrogen atoms together. This process is called nuclear fusion, and it's what makes stars shine.

One of the most exciting places in space is the black hole. A black hole is a region of space where gravity is so strong that nothing, not even light, can escape. Black holes are formed when massive stars collapse at the end of their lives. They can be incredibly massive, with some weighing as much as billions of suns. Black holes are fascinating because they challenge our understanding of the laws of physics.

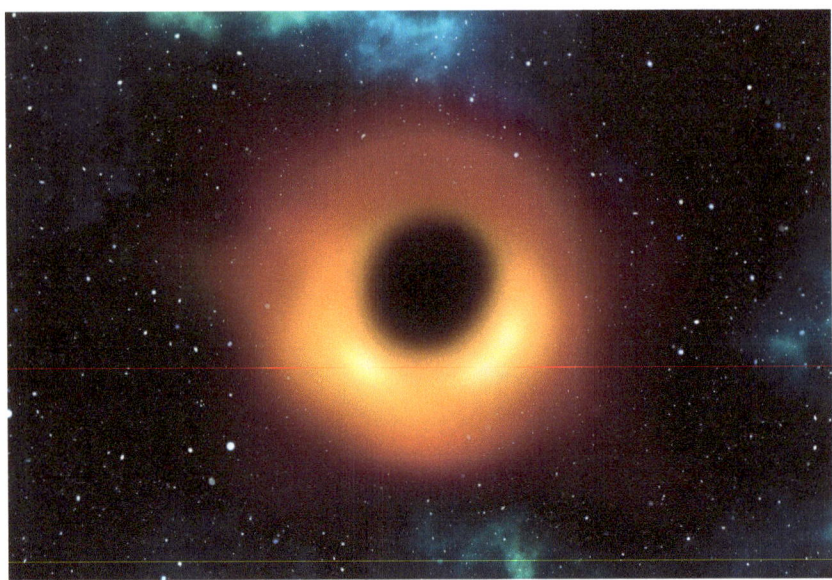

Have you ever heard of a galaxy before? It's a really big group of stars, planets, and other things in space, kind of like a giant city in space! Just like how there are different cities on Earth, there are also different galaxies in the universe.

The Milky Way is the name of the galaxy that Earth is a part of. It's really, really big, with billions and billions of stars inside of it. Imagine a big, giant room filled with so many toys that you couldn't even count them all - that's kind of what the Milky Way galaxy is like!

But the Milky Way is just one of many, many galaxies in the universe. There are so many galaxies that we can't even count them all! And just like how every city on Earth is different, every galaxy in the universe is also different. Some galaxies are big and spiral-shaped, while others are small and shaped like a ball.

Galaxies are really important because they help us understand more about the universe. By studying galaxies, scientists can learn more about how stars and planets are formed, and how the universe works. It's really fascinating to think about how big and complex our universe is!

So, galaxies are like giant cities in space that are made up of stars, planets, and other things. Earth is part of the Milky Way galaxy, which is just one of many, many galaxies in the universe. Studying galaxies can help us learn more about the universe and how it works!

Did you know that galaxies come in different shapes and sizes? Some galaxies are shaped like spirals, while others are shaped like balls or eggs. Just like how people come in different shapes and sizes, galaxies are also unique.

Inside of galaxies, there are many different things, like stars, planets, asteroids, and comets. There's even something called a black hole, which is a really, really, really big hole in space that sucks in everything around it, even light! Black holes are really interesting to scientists because they're so mysterious and different from anything else in the universe.

One really cool thing about galaxies is that they're constantly moving and changing. Sometimes, galaxies can collide with each other and form new galaxies. It's kind of like when two big trucks crash into each other and create a big mess - but in space, it creates something new and beautiful!

Studying galaxies can also help us learn more about how our own planet, Earth, was formed. It's like a big puzzle, and by studying different parts of the puzzle, we can learn more about how it all fits together.

So, galaxies are really fascinating and there's still so much we have to learn about them. By studying galaxies, we can learn more about the universe and the things that are outside of our planet. It's really amazing to think about how big and complex our universe is, and how much there is still to explore and discover!

The most distant galaxy ever observed is about 13 billion light-years away, which means that we're seeing it as it was when the universe was very young.

Space is an incredibly fascinating and mysterious place that has captured the imagination of humans for centuries. It's a vast, dark, and cold place that is home to countless planets, stars, galaxies, and other celestial bodies. By learning about space, we can expand our knowledge and understanding of the universe and our place in it. So, keep exploring and learning about space, and who knows what you might discover!

The study of space can open up a whole new world of knowledge and understanding for children. It can teach them about physics, astronomy, and other sciences, and also inspire them to ask questions about the world around them. Children can learn about the history of space exploration, including the first moon landing and the many missions that have explored other planets and their moons. They can also learn about the challenges that astronauts face when they travel to space, such as the lack of air, the extreme temperatures, and the lack of gravity. By learning about these challenges, children can develop an appreciation for the bravery and resilience of the men and women who have explored space.

In addition to learning about space, children can also participate in space-related activities, such as building and launching model rockets, stargazing, and even designing their own space missions. These activities can help children develop important skills such as problem-solving, critical thinking, and creativity. Moreover, they can also help children develop a sense of wonder and curiosity about the universe and inspire them to pursue careers in science and engineering.

Space is a fascinating and awe-inspiring subject that can captivate the minds of children and adults alike. By exploring the wonders of space, children can gain a better understanding of the world around them and develop a sense of curiosity and wonder that can last a lifetime. Whether through books, documentaries, or hands-on activities, there are countless ways to learn about space and inspire the next generation of explorers and scientists.

Let's learn about planets in space.

Planets are big, round objects that orbit, or go around, a star. There are eight planets in our solar system, which is the group of planets that go around our Sun. The eight planets are Mercury, Venus, Earth, Mars, Jupiter, Saturn, Uranus, and Neptune.

Each planet is different from the others and has its own unique features. For example, Earth has air to breathe and water, which makes it possible for us to live here. Mars is known for its red colour and has been explored by robots called rovers. Jupiter is the biggest planet in our solar system and is famous for its big red spot, which is a giant storm that has been raging for hundreds of years.

Planets can also be categorized into two groups: inner planets and outer planets. The inner planets are the ones closest to the Sun, and they are Mercury, Venus, Earth, and Mars. These planets are smaller and made mostly of rock and metal. The outer planets are the ones farther away from the Sun, and they are Jupiter, Saturn, Uranus, and Neptune. These planets are much bigger and made mostly of gas and ice.

One really cool thing about planets is that some of them have moons! A moon is a smaller object that orbits around a planet. Earth has one moon, but some planets like Jupiter have many moons.

Learning about planets is really fun because they come in all shapes, sizes, and colours. By studying planets, we can learn more about the different features of our solar system and the universe beyond.

Let's talk about the Sun!

The Sun is a really big ball of fire that gives off heat and light. It's actually a star, but it's the closest one to us, which is why we call it the Sun. It's so big that you could fit over a million Earths inside of it!

The Sun is really important because it gives us light and heat. Without the Sun, Earth would be a really cold and dark place. The light from the Sun helps plants grow, which is why we need it for food. The heat from the Sun also keeps us warm and helps us feel comfortable.

But we have to be careful around the Sun because it's very hot and bright. That's why we wear sunscreen and hats to protect ourselves when we play outside on sunny days. It's also why we should never look directly at the Sun because it can hurt our eyes.

The Sun is so big and powerful that it's always changing. Sometimes it has big explosions called solar flares, which can be dangerous for astronauts in space. But most of the time, the Sun is a really important and helpful part of our solar system.

So, the Sun is a really big and bright star that gives us light and heat. We need it to stay warm and for plants to grow. But we also have to be careful around it because it's very hot and bright.

Let's talk about stars in space.

Have you ever seen the stars in the night sky? They look like little twinkling lights, right? Well, those twinkling lights are actually really big balls of fire that are very, very far away from Earth. These balls of fire are called stars.

Just like how people come in different colours and sizes, stars also come in different colours and sizes. Some stars are really big, even bigger than the Sun, and some are really small, even smaller than Earth! Some stars are blue, some are white, some are yellow, and some are even red.

Stars are special because they give off light and heat. We can see the light from stars all the way from Earth, even though they're very far away. Stars are also important because they make up pictures in the night sky, like the Big Dipper, which looks like a big ladle or spoon.

But, did you know that stars are not just pretty to look at, they're also really important for us? That's because stars are what make up galaxies, which are big groups of stars, planets, and other things in space. Our planet, Earth, is part of the Milky Way galaxy, which is just one of many galaxies in the universe. We have already talked about this, remember ?

So, even though stars are far away and might look really small to us, they're actually really big and important for our universe!

Stars are really big balls of fire that give off heat and light. They come in different colours and sizes, and there are billions and billions of stars in the universe. Just like how the Sun is the closest star to us, other stars are the closest to other planets.

Stars can be really big, some are even hundreds of times bigger than the Sun! But they can also be really small, some are only a little bigger than Earth. Just like planets, stars also have different colours. Some stars are blue, some are white, some are yellow, and some are even red.

When you look up at the night sky, you might see stars twinkling. This happens because the light from the star has to travel through Earth's atmosphere before it gets to our eyes. The atmosphere can make the light bend and change, which makes the stars look like they're   twinkling.

Stars are really fascinating and there's still so much we have to learn about them. By studying stars, we can learn more about the universe and the things that are outside of our planet. It's really amazing to think about how big and complex our universe is!

Every planet in our solar system revolves, or orbits, around the Sun. The Sun is like the big boss of our solar system - everything revolves around it! When a planet orbits the Sun, it moves in a big circle or oval shape called an orbit.

The distance between the planet and the Sun determines how long it takes for the planet to complete one orbit. For example, Mercury is the closest planet to the Sun, so it only takes 88 Earth days for it to complete one orbit. But Neptune is the farthest planet from the Sun, so it takes 165 Earth years for it to complete one orbit!

The way a planet orbits the Sun also determines how much heat and light it gets. The closer a planet is to the Sun, the hotter it gets. That's why Mercury, the closest planet to the Sun, is very hot, while Neptune, the farthest planet from the Sun, is very cold.

Some planets in our solar system also have moons, which are like smaller versions of planets that orbit around the planet. It's kind of like a planet and its moon are best friends, and they travel together through space!

So, planets orbit, or revolve around the Sun, in a big circle or oval shape called an orbit. The distance between the planet and the Sun determines how long it takes for the planet to complete one orbit, and how much heat and light it gets. Some planets in our solar system also have moons that orbit around them. It's like a big dance party in space, with planets and moons all moving and grooving around the Sun!

Earth is one of the planets in our solar system that revolves around the Sun. It takes one year, or 365.25 days, for Earth to complete one orbit, or one trip around the Sun. That's why we have a birthday every year - it's like celebrating the anniversary of Earth's trip around the Sun!

As Earth orbits the Sun, it also spins on its axis, which is like an invisible line that runs from the North Pole to the South Pole. Day and night happen because Earth spins on its axis, which is like an imaginary line that runs from the North Pole to the South Pole. As Earth spins, the part of the planet facing the Sun has daylight, while the part facing away from the Sun has darkness.

When the side of Earth that you're standing on faces the Sun, it's daytime. The Sun gives off light and heat, and this makes it bright and warm outside. You can see everything around you, and it's a great time to play outside, go for a walk, or have a picnic.

But when the side of Earth that you're standing on faces away from the Sun, it's night time. The Sun is on the other side of the planet, so it's dark and cool outside. You can't see as well, but you can look up at the stars and the Moon, and it's a great time to snuggle up with a blanket and go to sleep.

Day and night happen all over Earth at different times because of the way Earth rotates. It takes Earth 24 hours to complete one full rotation on its axis, which is why we have day and night every day.

So, in summary, day and night happen because Earth spins on its axis. When the side of Earth that you're standing on faces the Sun, it's daytime, and when it faces away from the Sun, it's night time. It takes Earth 24 hours to complete one full rotation on its axis, which is why we have day and night every day.

The way that Earth revolves around the Sun also affects the seasons. Earth's axis is tilted, which means that during different times of the year, different parts of Earth get more or less direct sunlight. When the Northern Hemisphere is tilted toward the Sun, it's summer, and when it's tilted away from the Sun, it's winter. It's kind of like when you tilt a flashlight towards something - the part of the object that's facing the flashlight gets more light and heat, while the part that's facing away gets less.

Earth's orbit around the Sun also affects things like the tides, which are the rise and fall of the ocean. The Moon also orbits around Earth, and its gravity affects the tides. It's like the Moon and the ocean are best friends, and they work together to create the tides.

So, in summary, Earth revolves around the Sun in a big circle or oval shape called an orbit. It takes one year for Earth to complete one orbit. Earth also spins on its axis, which creates day and night, and its tilted axis affects the seasons. Earth's orbit around the Sun also affects things like the tides. It's like a big dance party in space, with Earth and the Sun and the Moon all moving and grooving together!

Let's talk about Mars !

Mars is a planet in our solar system that is known as the "Red Planet" because it has a reddish colour. It's the fourth planet from the Sun, and it's about half the size of Earth.

Mars is a really interesting planet because scientists think that it might have had water on its surface a long time ago, which means that it might have been able to support life! Scientists have sent robots called rovers to Mars to explore its surface and learn more about it.

One of the most famous rovers is called the Mars Curiosity Rover. It landed on Mars in 2012 and has been exploring ever since! The rover is about the size of a car and has lots of tools and instruments to help it study the rocks and soil on Mars. It takes pictures and sends them back to scientists on Earth, so we can see what Mars looks like up close!

Mars has a really thin atmosphere, which means that it doesn't have as much air as Earth does. This makes it a really cold and dry planet. The temperature on Mars can range from -195 degrees Fahrenheit at night to 70 degrees Fahrenheit during the day. Brrr, that's cold!

Mars also has two small moons called Phobos and Deimos. They're not very big, but they orbit around Mars just like the Moon orbits around Earth.

So, in summary, Mars is a planet in our solar system that is known as the "Red Planet" because it has a reddish color. It might have had water on its surface a long time ago, and scientists are studying it to learn more about it. Mars has a really thin atmosphere, which makes it a really cold and dry planet. It also has two small moons called Phobos and Deimos.

Can we go to the mars ?

As of now, humans have not yet travelled to Mars, but NASA (National Aeronautics and Space Administration) and other space agencies around the world are actively working on developing the technology and strategies needed to make it possible.

NASA has plans to send humans to Mars in the 2030s, and they are currently working on developing the necessary technology and infrastructure to make that happen. In fact, they are currently working on building a spacecraft called Orion, which will be used to transport humans to Mars.

There are still many challenges to overcome before humans can safely and effectively travel to Mars. For example, the trip to Mars can take several months, and the conditions on the planet are very different from those on Earth. However, scientists and engineers are working hard to develop new technologies and strategies to make human travel to Mars a reality.

In the meantime, we can continue to learn about Mars through the use of robotic explorers like the Mars rovers, which are able to collect data and images from the planet's surface.

Humans are interested in going to Mars for a number of reasons. For one, Mars is the planet that is most similar to Earth in our solar system. It has a similar length of day and night, and it has a rocky surface like Earth. This means that scientists believe that Mars might be able to support life, or may have been able to do so in the past.

Another reason why humans are interested in going to Mars is that it could help us learn more about the history of our own planet. Mars has a lot of similarities to Earth, but it also has some differences that could give us clues about the formation and evolution of our own planet.

Additionally, Mars has valuable resources such as water, which could be used to support human life on Mars or be brought back to Earth. Mars also has other resources such as metals and minerals that could be used for future space exploration.

Finally, going to Mars would be a major achievement for humanity. It would require the development of new technologies and the overcoming of many challenges, but it would also be an incredible accomplishment that could inspire future generations to explore and push the boundaries of what is possible.

Overall, there are many reasons why humans are interested in going to Mars, including the possibility of finding life, learning more about our own planet, accessing valuable resources, and achieving a major milestone in human exploration.

Scientists and engineers are trying to go to Mars because they want to learn more about the planet and see if it can support life. Mars is similar to Earth in some ways, like how it spins around the sun and has rocks and dirt on its surface. They also think Mars might have had life on it a long time ago! Going to Mars would be a really big challenge, but it would be very exciting for people because it would show what we can achieve and inspire us to explore more.

We need to go to Mars to learn more about the planet and see if it can support life. By exploring Mars, we can also learn more about our own planet and how it was formed. Additionally, Mars has valuable resources such as water and minerals that could be used to support future space exploration. Going to Mars would be a big accomplishment for humanity and could inspire future generations to explore and push the boundaries of what is possible.

Is earth becoming Dangerous ?

Earth is becoming dangerous due to a number of factors, including climate change, pollution, and deforestation. Climate change is caused by the increase in greenhouse gases in the Earth's atmosphere, which is primarily caused by human activities such as burning fossil fuels. This leads to rising temperatures, sea level rise, and extreme weather events like hurricanes and wildfires. Pollution, such as air pollution and water pollution, can have negative impacts on human health and the environment. Deforestation, or the clearing of forests for human activities such as agriculture and logging, can lead to the loss of habitats for animals and plants and contribute to climate change. It is important for us to take action to reduce our impact on the environment and protect the planet for future generations.

Earth is becoming dangerous because of things people are doing that harm the environment. For example, when people use cars and factories that create smoke and pollution, it makes the air and water dirty. This can make it harder for plants and animals to live and can make people sick too. When people cut down too many trees and forests, it can change the climate and make it hard for animals to find places to live. We need to take care of the Earth by not polluting, planting trees, and finding new ways to do things that are better for the planet.

Earth, our beautiful home !!

Earth is considered the best planet for life because it has all the things that living things need to survive, such as air, water, and a comfortable temperature range. Earth is also unique in having a variety of different ecosystems, like forests, oceans, and deserts, which support a huge diversity of plant and animal life. Additionally, Earth is located at the right distance from the sun, which provides just the right amount of warmth and light for life to thrive. These factors make Earth a very special and important place, and it's up to us to take care of it so that it can continue to be a wonderful home for all living things.

Our planet Earth is a truly special and unique place. It's the only planet that we know of that can support life as we know it. From the towering mountains to the depths of the oceans, Earth is full of wonders that fill us with awe and wonder. It's home to millions of different species, each one contributing to the incredible diversity of life on our planet. It's up to us to protect this amazing place and all the life that calls it home. We must take care of our planet, by reducing pollution, conserving resources, and preserving habitats. We owe it to future generations to ensure that they too can experience the beauty and wonder of our incredible planet. Let us all work together to take care of this fragile and precious home that we call Earth.

The importance of our planet Earth cannot be overstated. It provides us with everything we need to survive, from the air we breathe to the water we drink. The food we eat comes from the plants and animals that call Earth home. Our planet also provides us with natural resources like oil, gas, and minerals that we use to power our homes and cars, build our cities, and create the technology that makes our lives easier.

But the Earth is more than just a source of resources. It's a beautiful and dynamic planet that inspires us and fills us with wonder. The stunning landscapes and vibrant ecosystems remind us of the incredible power and diversity of nature. We're connected to the Earth in so many ways, from the air we breathe to the ground we walk on. It's up to us to take care of our planet and ensure that it remains a healthy and vibrant place for generations to come.

In recent years, we've seen the devastating effects of climate change and environmental destruction on our planet. We've witnessed the loss of biodiversity, the destruction of habitats, and the rise of extreme weather events like hurricanes and wildfires. It's clear that we need to take action now to protect our planet and preserve it for future generations.

Fortunately, there are many things we can do to help. We can reduce our carbon footprint by driving less and using renewable energy sources like solar and wind power. We can conserve resources by recycling, using less water, and choosing eco-friendly products. We can also support conservation efforts by volunteering, donating to environmental organizations, and speaking out for the protection of our planet.

Our planet Earth is an incredible place, and it's up to each and every one of us to take care of it. Let's work together to ensure that it remains a healthy and vibrant home for all living things.

www.ingramcontent.com/pod-product-compliance
Lightning Source LLC
Chambersburg PA
CBHW041945240526
45473CB00033B/575